Greg ha[...]

His men[...]

His men land in a bog.

Greg's best man is Big Red.

Greg grips his best man.

"Who are you?"

Greg asks him.

3

"Let me tell you!" said Big Red.

"I am Big Red from Smog.

You spell that S-M-O-G!"

4

Big Red said, "Help me!
Help me get on the log!"

Greg sets Big Red on the log.

"You are tops!" said Big Red.

"Let me be off to Smog!"

"Blast off to Smog!" said Greg.

The End